YOUR KNOWLEDGE HAS VALUE

On the Synthesis of the Distribution amongst the Integers of the Prime Number

William Fidler

Bibliographic information published by the German National Library:

The German National Library lists this publication in the National Bibliography; detailed bibliographic data are available on the Internet at http://dnb.dnb.de.

ISBN: 9783346835666
This book is also available as an ebook.

Print and binding: Books on Demand GmbH, Norderstedt, Germany
Printed on acid-free paper from responsible sources.

The present work has been carefully prepared. Nevertheless, authors and publishers do not incur liability for the correctness of information, notes, links and advice as well as any printing errors.

GRIN web shop: https://www.grin.com/document/1336494

On the synthesis of the distribution amongst the integers of the prime number counting function $\pi(k)$, viewed as a geometric object

W M Fidler

Abstract

A method is here devised in which the prime number counting function, $\pi(k)$, is viewed as having the properties of a staircase. Indeed, we use the terminology appropriate to a staircase to describe the parts of the distribution amongst the integers.

Proceeding from any prime number we generate strings of limited extent of consecutive integers which may (or may not) contain primes. It is arranged that the numbers at the boundaries of the strings are even, so that any primes can only be located within a string. The strings are generated with the aid of the iterative formula, $k_{n+1} = \left(\frac{k_1}{\ln k_1} \pm 1 \right) \ln k_n$, where k_1 is the even integer which precedes, (-) or follows, (+), any given prime number. It will be recognized that the first term within the bracket in this formula is the prime number counting function of C F Gauss. The iterative formula is very robust and converges rapidly when applied in either direction. In honour of Gauss we call the strings here generated, Gauss strings.

In conjunction with a variation of an accelerated version of trial division developed by the author [1] we can, in principle, construct the whole of the distribution of the prime number counting function without approximation.

It is shown also that the prime number counting function may be represented in number space by an infinite set of connected trapezia whose individual areas are numerically equal to the gap between the primes that are situated on the borders of any given trapezium.

Inhalt

Introduction..4

Analysis. 5

The Gauss string...8

The determination of the primality of a number 10

The simple geometry associated with a tread... 13

Discussion 15

References 16

Introduction

In 1859, Riemann published a short paper [2] which consisted of only six manuscript pages and in which he derived a formula for predicting the number of primes in any given range. In Zagier's inaugural lecture at the University of Bonn in May 1975, [3] a table of the prime numbers ranging up to 1,000,000,000 is shown (without attribution), where the predictions of Riemann's formula and the counted number, $\pi(k)$, of prime numbers in several ranges are compared. The agreement between these, whilst not perfect, is astonishing; even more so when it is realized that Riemann derived his result from Complex Analysis, which is a totally different branch of mathematics than Number Theory.

Ever since Riemann's work there have been numerous formulae derived for the predicting of the number of primes in any given range. Hardy, [4] showed that Riemann's formula is equivalent to Gram's series [5]. An extensive collection of formulae for $\pi(k)$, their ranges of applicability and their originators is given in WolframMathworld [6].

Analysis.

Fig1 shows our concept of the prime number counting function distribution. It is noted that the object resembles a staircase constructed by a demented carpenter. The treads of the staircase are of different length, whilst the risers are always of unit height.

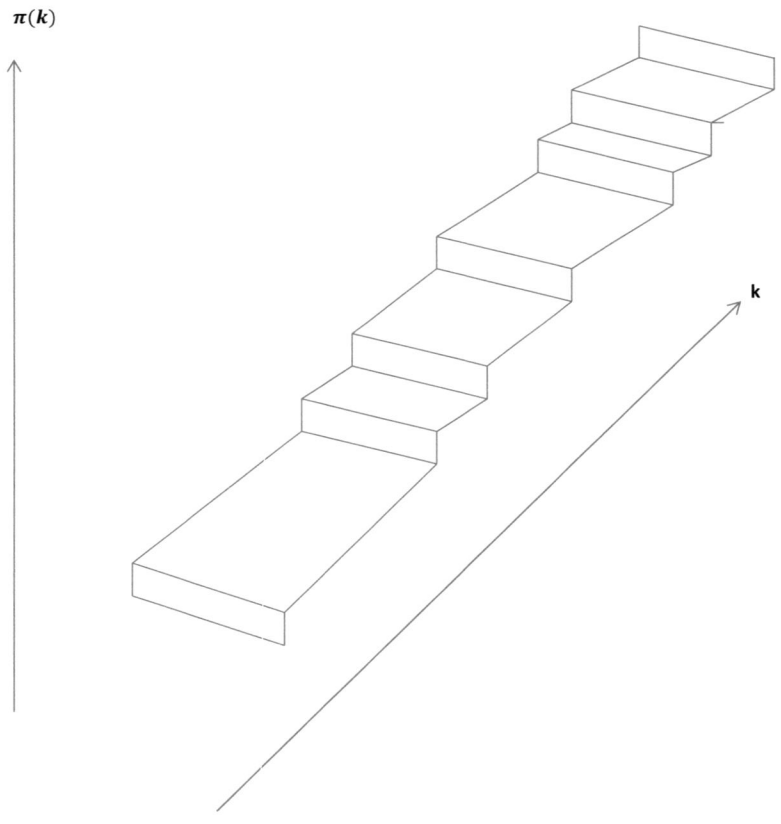

$\pi(k)$

k

Fig1

Fig2 shows the initial part of the prime number counting function, $\pi(k)$ presented in conventional form, and which, it may be noted may, without the vertical risers, be viewed as a version of a side elevation of an extension of **Fig1**.

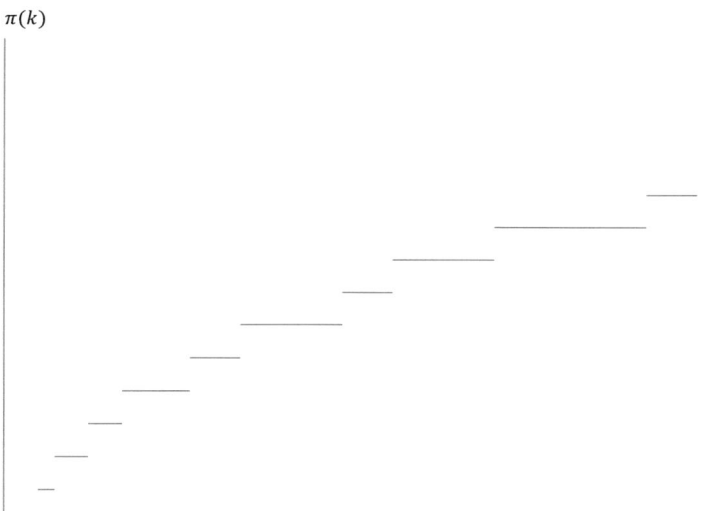

Fig2

In order to facilitate further discourse we present parts of a ∞ **x 15** matrix of the odd numbers developed in [1].

If we write out the odd numbers it is noted that every fourth number is divisible by 3 and hence cannot be prime. Although it is not the method presented here, it is seen that we may construct a series of 'cages' containing two numbers which may, or may not, be prime. Further, if we stack the odd number line as shown in the table below it is seen that we have formed a matrix having 15 columns and, assuming that the odd numbers extend to infinity, an infinite number of rows. The numbers shown underlined and in bold are prime numbers obtained using a prime number calculator available on Wikipaedia. Inspection of the array shows that, with the exception of columns which have a prime number as their first entry, there can never be prime numbers in columns 1, 2, 4, 7, 10, 12, and 13. Any number in the table (and the extension of the table to infinity) which is divisible by 3 can be rearranged to yield other numbers which cannot be prime; in the context of this work any rearranged number must not end in an even number and any rearranged number which ends in a 5 is a member of one of columns 2, 7, or 12. It should be noted that the 'spacing' between adjacent rows is 30, whilst that between adjacent columns is 2.

3	**5**	**7**	9	**11**	**13**	15	**17**	**19**	21	**23**	25	27	**29**	**31**
33	35	**37**	39	**41**	**43**	45	**47**	49	51	**53**	55	57	**59**	**61**
63	65	**67**	69	**71**	**73**	75	77	**79**	81	**83**	85	87	**89**	91
93	95	**97**	99	**101**	**103**	105	**107**	**109**	111	**113**	115	117	119	121
123	125	**127**	129	**131**	133	135	**137**	**139**	141	143	145	147	**149**	**151**
153	155	**157**	159	161	**163**	165	**167**	169	171	**173**	175	177	**179**	**181**
183	185	187	189	**191**	**193**	195	**197**	**199**	201	203	205	207	209	**211**
213	215	217	219	221	**223**	225	**227**	**229**	231	**233**	235	237	**239**	**241**
243	245	247	249	**251**	253	255	**257**	259	261	**263**	265	267	**269**	**271**
273	275	**277**	279	**281**	**283**	285	287	289	291	**293**	295	297	299	301
1	2	3	4	5	6	7	8	9	10	11	12	13	14	15

2013 2015 **2017** 2019 2021 2023 2025 **2027** **2029** 2031 2033 2035 2037 **2039** 2041

2043 2045 2047 2049 2051 **2053** 2055 2057 2059 2061 **2063** 2065 2067 **2069** 2071

2073 2075 2077 2079 **2081** **2083** 2085 **2087** **2089** 2091 2093 2095 2097 **2099** 2101

2103 2105 2107 2109 **2111** **2113** 2115 2117 2119 2121 2123 2125 2127 **2129** **2131**

2133 2135 **2137** 2139 **2141** **2143** 2145 2147 2149 2151 **2153** 2155 2157 2159 **2161**

2163 2165 2167 2169 2171 2173 2175 2177 **2179** 2181 2183 2185 2187 2189 2191

2193 2195 2197 2199 2201 **2203** 2205 **2207** 2209 2211 **2213** 2215 2217 2219 **2221**

2223 2225 2227 2229 2231 2233 2235 **2237** **2239** 2241 **2243** 2245 2247 2249 **2251**

2253 2255 2257 2259 2261 2263 2265 **2267** **2269** 2271 **2273** 2275 2277 2279 **2281**

2283 2285 **2287** 2289 2291 **2293** 2295 **2297** 2299 2301 2303 2305 2307 **2309** **2311**

2313 2315 2317 2319 2321 2323 2325 2327 2329 2331 **2333** 2335 2337 **2339** **2341**

2343 2345 **2347** 2349 **2351** 2353 2355 **2357** 2359 2361 2363 2365 2367 2369 **2371**

2373 2375 **2377** 2379 **2381** **2383** 2385 2387 **2389** 2391 **2393** 2395 2397 **2399** 2401

2403 2405 2407 2409 **2411** 2413 2415 **2417** 2419 2421 **2423** 2425 2427 2429 2431

2433 2435 **2437** 2439 **2441** 2443 2445 **2447** 2449 2451 2453 2455 2457 **2459** 2461

2463 2465 **2467** 2469 2471 **2473** 2475 **2477** 2479 2481 2483 2485 2487 2489 2491

2493 2495 2497 2499 2501 **2503** 2505 2507 2509 2511 2513 2515 2517 2519 **2521**

2523 2525 2527 2529 **2531** 2533 2535 2537 **2539** 2541 **2543** 2545 2547 **2549** **2551**

2553 2555 **2557** 2559 2561 2563 2565 2567 2569 2571 2573 2575 2577 **2579** 2581

2583 2585 2587 2589 **2591** **2593** 2595 2597 2599 2601 2603 2605 2607 **2609** 2611

1   2   3   4   5   6   7   8   9   10   11   12   13   14   15

The row of numbers immediately above here are the column numbers of the matrix.

The Gauss string

The concept of the Gauss string emanates from the assumption that if we choose, at random, an even number, say, k_1 then we can form Gauss' prime number counting function for that number, i.e. $\frac{k_1}{\ln k_1}$. If we increase this value by unity then we can create a range of consecutive numbers, one, or more of which might be prime. We call the range of numbers a Gauss string.

In the systematic construction of the prime number counting function we choose the even number alluded to above to be either the number which follows a prime number, or, the even one immediately preceding it, in the context of the direction in which we wish to generate the string.

To illustrate the method of synthesis consider the prime number **31**. Proceeding in the 'forward' direction we take $k_1 = 32$.

Hence we have, $k_{n+1} = \left[{k_1}/{\ln k_1} + 1 \right] \ln k_n = 10.233248 \ln k_n$, which iterates rapidly to a value of **36.9328**.

It was noted earlier that we begin and terminate the Gauss string with an even number and so we choose to terminate the above string at the number **38.**

The string then consists of the numbers **32,33,34,35,36,37,38.**

It is immediately obvious that the only number in the string that requires examination for primality is **37,** and, in this instance requires no complicated method to determine that **37 is** indeed a prime number.

We may then extend the last tread in **Fig2** as shown below:

31_____37

29____31

Fig3

It may be noted that this construction shows a typical feature of the method of synthesis, viz. the trailing prime of one tread, i.e. **31** is the leading prime of the next when proceeding in the 'forward' direction, whilst the converse is the case when proceeding in the 'backward' direction. We now proceed to determine the trailing prime of the next tread.

By the convention that we have adopted, in this instance k_1 equals **38** .

Hence we have: $k_{n+1} = \left[{k_1}/{\ln k_1} + 1 \right] \ln k_n = 11.44649 \ln k_n$, which iterates rapidly to a value of **43.0715**. As in the previous example we take the upper bound of the string to be an even number, and hence the string consists of the sequence, **38, 39, 40, 41, 42, 43, 44.**

Proceeding from the left the first number that we encounter that can be prime is **41**. By simple trial we find that **41** is indeed, a prime number. Hence, we may extend the distribution shown in **Fig3** as shown below.

37_____41

31_____37

Fig4

It may be noted that we have not included the number **43,** which is also prime, and it is here that we establish the rule that only the first prime in the string (proceeding in either direction from the 'reference' prime) is considered in its contribution to the value of the prime number counting function.

We then take k_1 equal to **42,** and then proceed to generate a new string from:

$$k_{n+1} = \left[{}^{k_1}\!/_{\ln k_1} + 1 \right] \ln k_n = 12.23694 \ln k_n.$$ This iterates to **47.1539,** and so, the string, in this instance is **42, 43, 44, 45, 46, 47, 48.** As before we proceed from the left and find that the first number that can be prime is **43.** Hence, we may extend **Fig4** as follows:

```
                                                    41       43
                                      37            41
                        31                          37
            29     31
```

Fig5

It will be seen that if, starting from the beginning of the prime numbers, we number the treads of the staircase, then each number is equal to the magnitude of the prime number counting function.

The determination of the primality of a number

The utility of the Gauss string has been demonstrated, in that, in a systematic manner it presents a limited range of numbers which may contain a prime, or prime numbers. Many of the members of the string may be eliminated by simple inspection, or by the means noted earlier in this work, but there remains the problem of determining the primality, or otherwise of those remaining numbers whose nature cannot be so readily decided.

We now present a variation of the accelerated method of trial division developed in [1].

Consider any number, **N.** From [1] we may derive the following expression:

$$I = \frac{[(N-1) - 2n]}{2n + 1} \quad \text{------------------ (1)}$$

Here, **n** is the sequence; 1 ,2 ,3 ,4 ,5 --------------∞.

The above equation may be written: $I = {N}/{2n+1} - 1$. However, the author regards the numerical results of (1) to be aesthetically pleasing and hence the latter equation is not used.

If $2n + 1$ is a factor of **N** then **I** will be an integer. Hence, in order to determine the nature of **N** all that is required is that we substitute values of **n** in the sequence shown above. The range of **n** for any given **N** is obtained by taking the square root of **N**, rounding it up to the next odd number and then determining that value from the formula, $2n + 1 = \sqrt{N}^{\uparrow}$,

The process is best illustrated by example.

Let N = 59; therefore, $\hat{n} = \frac{9-1}{2} = 4$. We may then write, $I = {[58 - 2n]}/{2n+1}$.

Consider the following table.

n	1	2	3	4
I	56/3	54/5	52/7	50/9

None of the **I** is an integer, hence **59** is prime.

Let N = 187; therefore, $\hat{n} = \frac{15-1}{2} = 7$. We may then write, $I = {[186 - 2n]}/{2n+1}$.

n	1	2	3	4	5	6	7
I	184/3	182/5	180/7	178/9	176/11	174/13	172/15

By simple inspection we see that at **n = 5, I** is equal to **16**, hence **11** is a factor of **187** and so **187** is not prime.

From the above examples it will be obvious that this process is particularly simple, in that, after forming **I** for **n = 1,** we may form all of the other **I** by the simple process of decreasing the numerator of the fraction by **2,** whilst increasing the denominator by **2.**

Hence we have a simple process for examining any number of interest within the Gauss string.

It should be noted that we are not limited to proceeding in the structured manner which has been demonstrated in **Fig2**. Indeed, we may choose any even number and generate the Gauss string emanating therefrom. Let us illustrate this by choosing, at random, the value **2526** from the matrix shown before.

Hence, we have: $k_{n+1} = \left[{k_1}/{\ln k_1} + 1 \right] \ln k_n = 323.4244 \ln k_n$, which iterates to **2534.982.** The string is then: **2526, (2527), 2528, 2529, 2530, (2531), 2532, (2533), 2534, 2535, 2536.** We may now examine the numbers which have been underlined and bracketed for all of the others cannot be prime.

Let $N = 2527$; therefore, $\hat{n} = \frac{51-1}{2} = 25$. We may then write, $I = \frac{[2526 - 2n]}{2n + 1}$, and construct the associated table:

n	1	2	3	4	5
I	2524/3	2522/5	<u>2520/7</u>	2518/9	2516/11

Whilst the table should have extended to **n = 25,** its truncation demonstrates what we conjecture to be a characteristic feature of this type of computation, viz. that if the number under examination is not prime, then we may encounter one of its factors early in the sequence, and, this is the case in this instance for **2520/7 = 360,** hence **7** is a factor of **2527** and so **2527 is** not a prime number; we now move to examine **2531,** which, when examined by the same approach, (except we must extend the table to **n = 25)** shows that there are no factors, thus **2531** is a prime number. Further, we consider that the process used here is more computationally-efficient than simple trial-division, in that, as shown above, the numerator of the fraction is continually decreasing, whilst the denominator is continually increasing, and so the process of division, which is nothing more than repeated subtraction will accelerate as we proceed from left to right along the second row in the above tables.

We complete this section by showing that we may use the Gauss string in the 'backward' direction.

It has been shown above that the number **2531** is prime. Let us take **2530** to be the initial number in the Gauss string.

Hence, we have: $k_{n+1} = \left[\frac{k_1}{\ln k_1} - 1 \right] \ln k_n = 321.87 \ln k_n$, which iterates to **2521.018.**

The string is then: **2530, 2529, 2528, 2527, 2526, 2525, 2524, 2523, 2522, (2521), 2520.**

Having shown that **2527** is not a prime number, the only number in the string which can be prime is **2521.** By exactly the same procedure as before we can show that **2521** is indeed, a prime number.

The simple geometry associated with a tread

Fig6 shows a tread and riser with the leading edge of the tread connected by a construction line of length, **L** to the leading edge of the next higher tread.

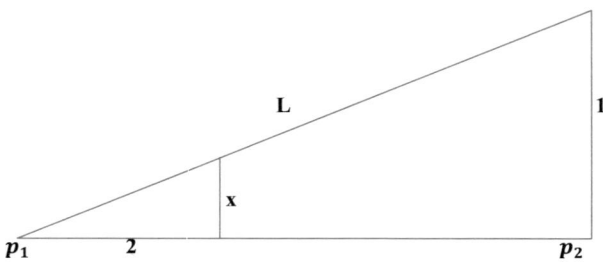

Fig6

Let θ be the angle between the line **L** and the line p_1------ p_2, and since we are considering a tread, then the 'p' are prime numbers. The line of length, **x**, is located at the next odd number after, p_1.

Now, $\tan \theta = {}^x/_2 = {}^1/_{(p_2 - p_1)}$, hence, $p_2 - p_1 = {}^2/_x$; but, ${}^2/_x$ must be an even integer. From this it follows that we must set, $x = {}^1/_n$, where, n = **1, 2, 3, 4, -------∞.**

Hence, $p_2 - p_1 = 2n$, which is the gap between the primes and, $\tan \theta = {}^1/_{2n}$. The triangle may now be shown with one side equal to unity whilst the other two sides are simple functions of **n.**

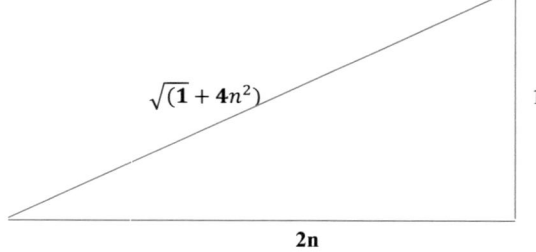

Fig7

Further, it is easy to show that the product of the area of the triangle, **A** and $\tan \theta$ are given by:

$$A \tan \theta = \frac{1}{2} \text{----------------------- (2).}$$

This is the classic equation of a rectangular hyperbola. In addition, the smallest triangle that may be generated corresponds to a set of prime twins. In this case $A = 1$ and, $\tan \theta = \frac{1}{2}$.

It is noted that there cannot be two of these smallest triangles in this representation of the prime number counting function in the whole of the range of the integers which are adjacent to each other, since we have shown in [1] that the only three consecutive odd numbers which can be prime are **3, 5** and **7** and such a combination exists nowhere else; also, θ can never be greater than $\tan^{-1} \frac{1}{2}$, whilst **A** increases without limit. A moment's reflection shows that both **A** and $\tan \theta$ are quantised, which is characteristic of a discontinuous function such as the prime number counting function.

We may extend the geometry of the foregoing, as follows:

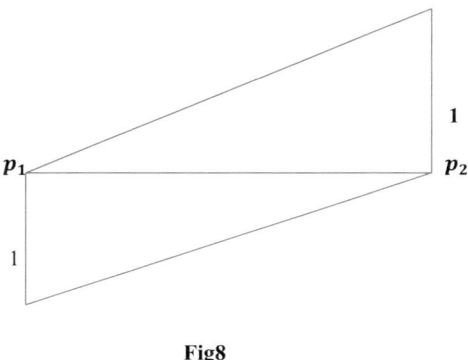

Fig8

The area of the trapezium above is $p_2 - p_1 = 2n$, but, this is the gap between the primes. Hence, since the above construction could be made anywhere in the range of the integers, we may state that the prime number counting function can be represented in number space by an infinite set of connected trapezia, each of which has an area which is numerically-equivalent to the gap between the primes.

14

Discussion

Given the comprehensive treatment of the foregoing analysis, there is little further to add to the discussion of the work.

There is however, one criticism that the author would make regarding the efforts to fit the discontinuous prime number counting function by a series of continuous functions. Such functions may be evaluated at any point and so there must be an infinite number of regions within the range of these functions, in which the values of the function are utterly meaningless, in that they cannot correspond to the value of the prime number counting function for such is not even defined within the region. Using our staircase terminology we may state this succinctly by the phrase 'there are no values of the prime number counting function within the risers'.

W M Fidler

March 2023

References

[1] On the determination of the primality of a number by the use of an accelerated version of trial division.

   W M Fidler. May 2021.

    ISBN 9783346493002.

[2] On the Number of Primes less than a Given Magnitude.

    G F B Riemann

   The Monthly Notices of the Berlin Academy.

    November 1859.

[3] The first 50 million prime numbers.

    Don Zagier

    Inaugural Lecture.

    Bonn University, May 1975.

[4] Ramanujan: twelve Lectures on Subjects suggested by his work.

    G H Hardy.

    3rd Ed. New York: Chelsea, pp24-25.

                1999.

[5] Undersogelser angaende Maengden af Primtal under en given Graense.

    J P Gram

    Kong. Dansk

    Videnskab. Selsk. Skr(VI) 2, pp183-308

                1884.

[6] WolframMathworld.

YOUR KNOWLEDGE HAS VALUE